Transportation Conformity:
A Basic Guide for State and Local Officials

A Message To The Reader

The planning provisions of the Intermodal Surface Transportation Efficiency Act (ISTEA) of 1991 and the transportation provisions of the Clean Air Act Amendments (CAA) of 1990 define the framework for the effective integration of transportation and air quality planning. The goal of transportation conformity, which is a key element in the planning process, is to ensure that air quality considerations are an integral part of transportation decisions.

Federal, State, and local agencies have been working hard to integrate the transportation conformity requirements as spelled out in the November 1993 conformity regulation, which resulted from a collaborative effort by the Environmental Protection Agency (EPA), the Federal Highway Administration (FHWA), and the Federal Transit Administration (FTA) to implement the CAA's transportation provisions.

FHWA, FTA, and EPA have been working since 1993 to identify new ways our agencies can work together to provide additional flexibility and reduce or eliminate conformity implementation problems. We are confident that the new amendments to the conformity rule will assist transportation and air quality agencies in more easily meeting their transportation and air quality goals. The new amendments provide additional flexibility through greater emphasis on State and local decisions.

Experience has also demonstrated the need to do a better job in explaining conformity to State and local officials who make decisions and trade-offs on transportation investments, including the impact of those decisions on air quality. This guide explains the rationale for linking transportation and air quality planning and the basics of transportation conformity. It is intended to help officials, policy makers, and others understand transportation conformity and their respective roles and responsibilities in the conformity process.

Today's transportation decisions will impact mobility and air quality in the coming decades and affect the quality of life for the next generation of Americans. This publication will assist officials by guiding them through the conformity process toward decisions that reflect a balancing of the important national goals of clean air and improved mobility.

Jane F. Garvey
Acting Administrator
Federal Highway Administration

Gordon J. Linton
Administrator
Federal Transit Administration

Table of Contents

A Message to the Reader iii
Executive Summary vii
Part I: The Basics of Transportation Conformity 1
   Background 1
   What Is Transportation Conformity? 1
   What Actions Are Subject to Transportation Conformity? 1
   Who Makes Conformity Determinations? 2
   What Is the Frequency of Conformity Determinations? 2
   The Key Components of a Conformity Determination 2
      CAA Requirements 3
      ISTEA Planning Requirements 4
   Consequences of a Failure to Make a Conformity Determination ... 4
      Lapsing 5
      Trade-offs 5

Part II: Roles and Responsibilities in the Conformity Determination 7
   Interagency Consultation 7
   Public Participation: Access for Stakeholders and Citizens 7

Conclusion 9

Appendices
   Appendix A: Options to Reduce Emissions from Motor Vehicles 11
   Appendix B: Health Impacts of Pollutants 15
   Appendix C: Resource Agencies Contact List 17

Glossary 19

Executive Summary

The CAA and ISTEA provisions are intended to ensure that integrated transportation and air quality planning occurs in the areas designated by EPA as nonattainment or maintenance areas.[1] The transportation conformity process establishes the major connection between transportation planning and emission reductions from transportation sources. To fully appreciate this connection, it is important to know about the key linkages between the CAA and ISTEA.

This guide discusses the basic provisions of the conformity process, including the following:

- A description of actions subject to conformity.
- Frequency of conformity determinations.
- Key components of a conformity determination.
- Consequences of a failure to make a conformity determination.
- Roles and responsibilities of public agency staff, management, policy officials, and decision makers in the conformity process.

ISTEA Links Conformity to Funding

Transportation conformity is an analytical process required of Metropolitan Planning Organizations (MPOs) and, in some cases, States, pursuant to the Clean Air Act Amendments (CAA) of 1990. ISTEA links compliance with the conformity requirements to continued FHWA and FTA funding of transportation plans, programs, and projects.

> *Under ISTEA's metropolitan planning requirements, projects cannot be approved, funded, advanced through the planning process, or implemented unless those projects are in a fiscally constrained and conforming transportation plan and transportation improvement program (TIP).*

ISTEA Links Conformity to Planning

In the past, a full assessment of the air quality impacts of transportation investments had not always been evident in State and MPO transportation plans, programs, and projects. Under the CAA, States and MPOs must demonstrate, through the conformity process, that the transportation investments, strategies, and programs they choose, taken as a whole, have air quality impacts consistent with those contained in State plans (State Implementation Plan (SIP)) for achieving the National Ambient Air Quality Standards (NAAQS-hereafter referred to also as Standards) and that emissions do not exceed the SIP targets for emissions from mobile sources.

Options and Trade-offs Between Emission Reduction Strategies

State and local elected officials are responsible for deciding what transportation investments the State and/or MPOs will make and how the State will attain the Standards for various pollutants. Officials need to be aware of the options available and the trade-offs their decisions entail so they can balance the need for transportation investment with the need to attain the Standards and thus achieve healthful air.

The options available to reduce emissions from motor vehicles and the health impacts of pollutants related to motor vehicle travel are discussed in the appendices.

> *It is hoped that by understanding the full array of options to reduce emissions, elected officials will appreciate the importance of transportation conformity as a tool that serves the twin goals of clean air and improved mobility.*

---

[1] The National Highway System Designation Act of 1995 restricted the application of the conformity requirements to nonattainment and maintenance areas.

CAA Linkages to Transportation Planning

The CAA requires that each State develop a SIP that addresses each pollutant for which the State fails to meet the Standards and indicates how the State intends to meet the Standards on schedules prescribed in the CAA. The Standards are usually expressed in terms of parts of pollutant per million molecules of ambient air (ppm) and vary by pollutant type. The key transportation-related pollutants are ozone, carbon monoxide, and particulates. A region can be designated as a nonattainment or maintenance area for one or more pollutants and can have different designations based upon the severity of violations for each pollutant.

Emissions are generally classified in one of three categories: stationary sources, area sources, and mobile sources.

- Stationary sources are relatively large, fixed sources of emissions such as power plants, chemical process industries, and petroleum refineries.
- Area sources are small, stationary and non-transportation sources that may collectively contribute to air pollution (e.g., dry cleaners, bakeries, etc.).
- Mobile sources include on-road sources such as cars, trucks, and buses, and off-road sources such as trains, ships, boats, airplanes, lawnmowers, and construction equipment.

There are also natural emissions, called biogenic, which come from the life processes of plants and animals; these are uncontrollable but also contribute to the formation of ozone.

Transportation officials are responsible for finding ways to reduce emissions from on-road mobile sources.

The SIP assigns emission reduction targets to each source category. For the mobile source category, the emission reduction target is further refined into a regulatory limit on emissions, referred to as a "budget," as discussed below.

Emission reduction targets for mobile sources can be achieved through programs that address vehicle emissions (e.g., the use of reformulated gasoline, implementation of Inspection and Maintenance (I & M) programs); by changing how we travel (e.g., ridesharing or use of transit) or from transportation investments to reduce congestion (e.g., signal synchronization programs).

Challenges for Transportation Officials

Transportation decision makers face two distinct challenges.

- First, they must work with State air quality officials to assess trade-offs between mobile and stationary source emission reduction programs and adopt a specific set of SIP strategies to enable them to demonstrate that reductions needed to reach attainment can be achieved. Strategies vary in cost, effectiveness, and ease of implementation, and a host of factors need to be considered in deciding which strategies to select.
- Second, they need to adopt a transportation plan and Transportation Improvement Program (TIP) that will enable the nonattainment area to stay within the SIP-adopted mobile source budget. If the State and MPO cannot demonstrate that the selected strategies and their resulting emission reductions are consistent with the SIP, then transportation projects and programs can be halted. (See section on consequences of failing to make a conformity determination.)

The mobile source "emissions budget" included in the SIP represents the highest level (ceiling) of emissions allowed from all projects included in the 20-year regional transportation plan and TIP.

Even after an area attains the Standards, it cannot exceed this ceiling on emissions and must identify ways to offset any future emissions increases due to population and employment growth, and expected increases in vehicle miles traveled. A limited number of nonattainment areas are exempted from the requirement to have a budget.

The State environmental agency assigns emission reductions to all pollution sources. Transportation officials should participate in decision making on the SIP and allocation of reductions to source categories. The State must seek EPA approval to revise SIP strategies and/or if it cannot meet its commitments to reduce emissions from EPA approved SIP strategies. It is important that the level of emission reductions assigned to each of the sources of pollution can be achieved through the implementation of the selected strategies.

Interagency Consultation Requirements and Public Participation

The success of the conformity process depends upon State and local transportation and air quality agencies working together.

The conformity regulation requires that State and local agencies establish formal procedures to ensure interagency coordination on critical issues. Public participation in transportation and air quality planning is also discussed in this guide. In the past five years, much progress has been made in integrating transportation and air quality planning. This guide is intended to further facilitate that coordination and integration.

Part One: The Basics of Transportation Conformity

Background

The concept of transportation conformity was introduced in the Clean Air Act of 1977, which included a provision to ensure that transportation investments conform to the SIP for meeting the NAAQS. Conformity requirements were made substantially more rigorous in the CAA of 1990, and the transportation conformity regulation that details implementation of the new requirements was issued in November 1993.[2] The regulation details the process for transportation agencies to demonstrate and ensure emission reductions from transportation sources of air pollution.

What Is Transportation Conformity?

> *Conformity is a way to ensure that Federal funding and approval are given to those transportation activities that are consistent with air quality goals.*

It ensures that these transportation activities do not worsen air quality or interfere with the "purpose" of the SIP, which is to meet the NAAQS [3]. Meeting the Standards often requires emission reductions from mobile sources.

According to the CAA, transportation plans, programs, and projects cannot

- create new NAAQS violations,
- increase the frequency or severity of existing NAAQS violations, or
- delay attainment of the NAAQS.

What Actions are Subject to Transportation Conformity?

Transportation Plans and Transportation Improvement Programs

The CAA requires that transportation plans, programs, and projects in nonattainment or maintenance areas that are funded *or approved* by the FHWA or FTA be in conformity with the SIPs through the process described in the EPA's transportation conformity regulation (See Figure 1).

Under ISTEA, MPOs must have transportation plans in place that present a 20-year perspective on transportation investments for their region. The transportation improvement program (TIP) is a multi-year prioritized list of projects (3 years at a minimum) proposed to be funded or approved by FHWA or FTA.

> *The TIP must be consistent with the conforming transportation plan, and the TIP must be found to conform to the SIP. Specifically, the transportation plan and TIP must result in emissions consistent with those allowed in the SIP.*

Regionally significant [4] transportation projects that are not funded *or approved* by FHWA and/or FTA, but which are sponsored by recipients of FHWA/FTA funds, must also be included in the Plan and TIP conformity analysis. In rural nonattainment or maintenance areas, the State must ensure that regionally significant projects conform to the SIP.

---

[2] U.S. Environmental Protection Agency, (1993) *Criteria and Procedures for Determining Conformity to State or Federal Implementation Plans of Transportation Plans, Programs, and Projects Funded or Approved Under Title 23 U.S.C. or the Federal Transit Act,* Title 40 C.F.R., Parts 51 and 93, November 24, 1993, as amended in August and November 1995.

[3] Any activity (funded, approved, permitted, etc.) undertaken by Federal agencies, other than the FHWA and the FTA, is governed by separate conformity regulations. Project level conformity is also required and, under certain circumstances localized emissions analysis is required. This requirement is discussed later in this document.

[4] "Regionally significant project" means a project that is on a facility that serves regional transportation needs and would normally be included in the modeling of a metropolitan area's transportation network. This includes, as a minimum, all principal arterial highways and all fixed guide-way transit facilities that offer a significant alternative to regional highway travel.

Project Level Conformity

FHWA/FTA projects must be found to conform before they are adopted, accepted, approved, or funded. With some exemptions[5] (e.g. safety, landscaping, and other projects with neutral or minimal emissions impacts), transportation projects must conform to the following criteria:

- They must come from a conforming transportation plan and TIP.
- The design concept and scope of the project that was in place at the time of the conformity finding must be maintained through implementation.
- The project design concept and scope must be sufficiently defined to determine emissions at the time of the conformity determination.

If a project does not meet the above three criteria, its emissions, when considered with the emissions projected for the conforming transportation plan and program, cannot cause the plan and program to exceed the emissions budget in the SIP. Areas that have carbon monoxide (CO) or particulate matter (PM_{10}) problems must also show that new localized violations of those pollutants will not result from project implementation.

Who Makes Conformity Determinations?

The MPO and U.S. DOT have a responsibility to ensure that the transportation plan and program within the metropolitan planning boundaries conform to the SIP. In metropolitan areas, the policy board of each MPO must formally make a conformity determination on its transportation plan and TIP prior to submitting them to the U.S. DOT for review and approval. Conformity determinations for projects outside of these boundaries are the responsibility of the U.S. DOT and the project sponsor, which usually is the State DOT.

What Is the Frequency of Conformity Determinations?

Conformity determinations must be made at least every three years, or as changes are made to plans, TIPs, or projects. Certain events, such as SIP revisions that establish or revise a transportation-related emissions budget, or add or delete transportation control measures (TCMs), may also trigger new conformity determinations.

The Key Components of a Conformity Determination

> *The key components of the conformity determination include regional emissions analysis, project level analysis, and, if TCMs are part of the attainment demonstration, the providing of evidence of the timely implementation of TCMs.*

The foundation upon which a conformity determination is based is the motor vehicle emissions budget in the latest submitted or approved SIP. Once established and approved by EPA, it is this budget upon which regional analysis is based. The regional analysis must comply with specific modeling requirements included in the regulation.

In areas where budgets are not required, emission reduction tests are used[6]. In addition, compliance with ISTEA's planning requirements is integral to making a conformity determination. These requirements include demonstration of a fiscally-constrained plan and TIP, and consideration of planning factors.

---

[5] 40 CFR Parts 51 and 93, *Federal Register*, November 24, 1993, page 63214.

[6] Emission reduction tests (also called "build/no-build" tests) are a comparison of the emissions impacts of implementing the proposed plan and/or TIP to the emissions impacts of not implementing the plan and/or TIP.

CAA Requirements
Regional Emissions Analysis

Consistency of transportation plans and TIPs with the motor vehicle emissions budget in the SIP is achieved by performing a regional emissions analysis. This analysis includes emissions resulting from

- the entire transportation network in the nonattainment or maintenance area,
- all proposed regionally significant projects,
- the effects of any emission control program(s) already adopted by the enforcing jurisdiction (e.g., vehicle inspection and maintenance programs, reformulated gasoline programs).

Timely Implementation of TCMs

In areas where TCMs are included in the SIP, the MPO or State must ensure that all TCMs have funding priority consistent with the SIP schedule for implementation. This provision is incorporated into the conformity process partly to insure that TCMs are not postponed due to lack of a funding commitment. This can be a useful tool in reinforcing the linkages between SIPs and transportation plans and TIPs, and may require local, regional, and State transportation officials to make investment trade-offs between projects to ensure TCMs are implemented.

Project Level Emissions Analysis

Project level emissions analysis (hot spot analysis) applies to carbon monoxide and PM_{10} concentrations and is based on quantitative analysis using applicable air quality models. Areas can establish their own thresholds for quantitative analysis with EPA approval. In some cases, qualitative analysis can be used. Quantitative analysis is required for the types of projects listed below:[7]

- Projects that are identified in the applicable SIP as sites of violation or possible violation.
- Any project affecting one or more of the top three intersections with the highest traffic volumes, or worst level of service (LOS) in a nonattainment or maintenance area.
- Projects affecting intersections that are at level of service (LOS) D, E, or F.

Emissions Budgets

An emissions budget for motor vehicles is the total of all motor vehicle emissions identified in the SIP that an area can produce and still achieve the SIP's purpose, which is to demonstrate attainment of the NAAQS or, in some cases, demonstrate maintenance.

In effect, motor vehicle emissions budgets are quantification of the "carrying capacity" of the region for each pollutant type and are reduced gradually over time as the area nears attainment. Thus, the carrying capacity equals the budget.

Budgets are developed based upon the emissions inventory in the SIP and reflect effects of control measures included in the SIP. Motor vehicle emissions inventories are based upon the number of vehicles in the region, their age, the rate of fleet turnover to newer and cleaner vehicles, seasonal temperatures in the region, and other factors. The motor vehicle emissions budget in the SIP acts as a ceiling on the transportation plan and TIP emissions.

---

[7] PM_{10} hotspot analysis will not be required until EPA releases modeling guidance on this issue and announces it in the *Federal Register*.

Modeling

The conformity regulation specifies detailed modeling requirements. The CAA requires that the latest planning assumptions be used in the conformity analysis. In some cases, these are not the same assumptions that were used in the SIP development process.

- Travel demand models must be used to estimate how much travel will occur in the region based on travel characteristics and growth assumptions.
- Emissions models must be used to estimate regional emissions (these estimates are derived from grams of pollutant per mile traveled) and are based upon the output of the travel demand model.
- Air quality dispersion models are used to evaluate localized impacts (project level impacts) of carbon monoxide emissions.

These requirements have presented challenges to transportation and air quality agencies, and promising work is underway to improve modeling capabilities for both travel demand and emissions estimation.

ISTEA's Planning Requirements

ISTEA's planning framework, in addition to reinforcing CAA transportation requirements, incorporates two additional provisions that support the integration of air quality and transportation planning. MPO plans must demonstrate fiscal constraint and consideration of ISTEA's planning factors.

Fiscal Constraint Requirements

In the first two years of the TIP, only projects that can be implemented with funds that are available or committed may be included. This fiscal constraint requirement helps to ensure that the projects included in transportation plans and TIPs, taken as a whole, are funded and implemented.

> *Under the provisions of ISTEA, non-attainment and maintenance areas must show in their regional transportation plan and in the TIP that funds are reasonably expected to be available to maintain and operate the existing transportation system and to implement new projects and programs on schedule.*

This provision is incorporated into the conformity process in part to insure conformity findings are based on realistic plans and programs, and that TCMs and other projects which may be beneficial to air quality are given priority.

Consideration of Planning Factors

ISTEA requires that MPOs consider 16 different factors in their transportation plans to ensure that a comprehensive set of potential impacts are considered prior to making investment decisions. The factors include environmental consequences of decisions—including land use, energy, and air quality impacts. The requirement reinforces the importance of understanding linkages between transportation and air quality planning.

Consequences of a Failure to Make a Conformity Determination

The conformity determination is a result of analyses undertaken to identify both the projected regional emissions impacts of plans and programs and the project level impacts of individual projects. If a transportation plan, program, or project does not meet conformity requirements, transportation officials can choose from the following options:

- Modify the plan, program, or project to offset the expected emissions.
- Work with the appropriate State agency to modify the SIP to offset the plan, program, or project emissions.

If the above is not accomplished, projects cannot advance. This can affect transit as well as highway projects. There are exceptions, however, and they include specific categories of projects such as purchase of rail cars and highway safety projects, projects which have completed the conformity and NEPA processes, and some projects that are not funded with FHWA or FTA administered funds.[8]

Lapsing

If a conformity determination cannot be made within specified time frames after amending the SIP, or three years passes since the last conformity determination, the determination lapses and no new projects may advance until a new determination for the plan and TIP can be made.

Trade-offs

Given the possible consequences of a failure to make a conformity finding, elected officials and decision makers need to be prepared to make difficult choices should they be faced with this situation.

In considering trade-offs, an understanding of the options to reducing mobile sources of emissions is important and is discussed in Appendix A.

---

[8] NEPA means National Environmental Policy Act of 1969, as amended (42 U.S.C. §321).

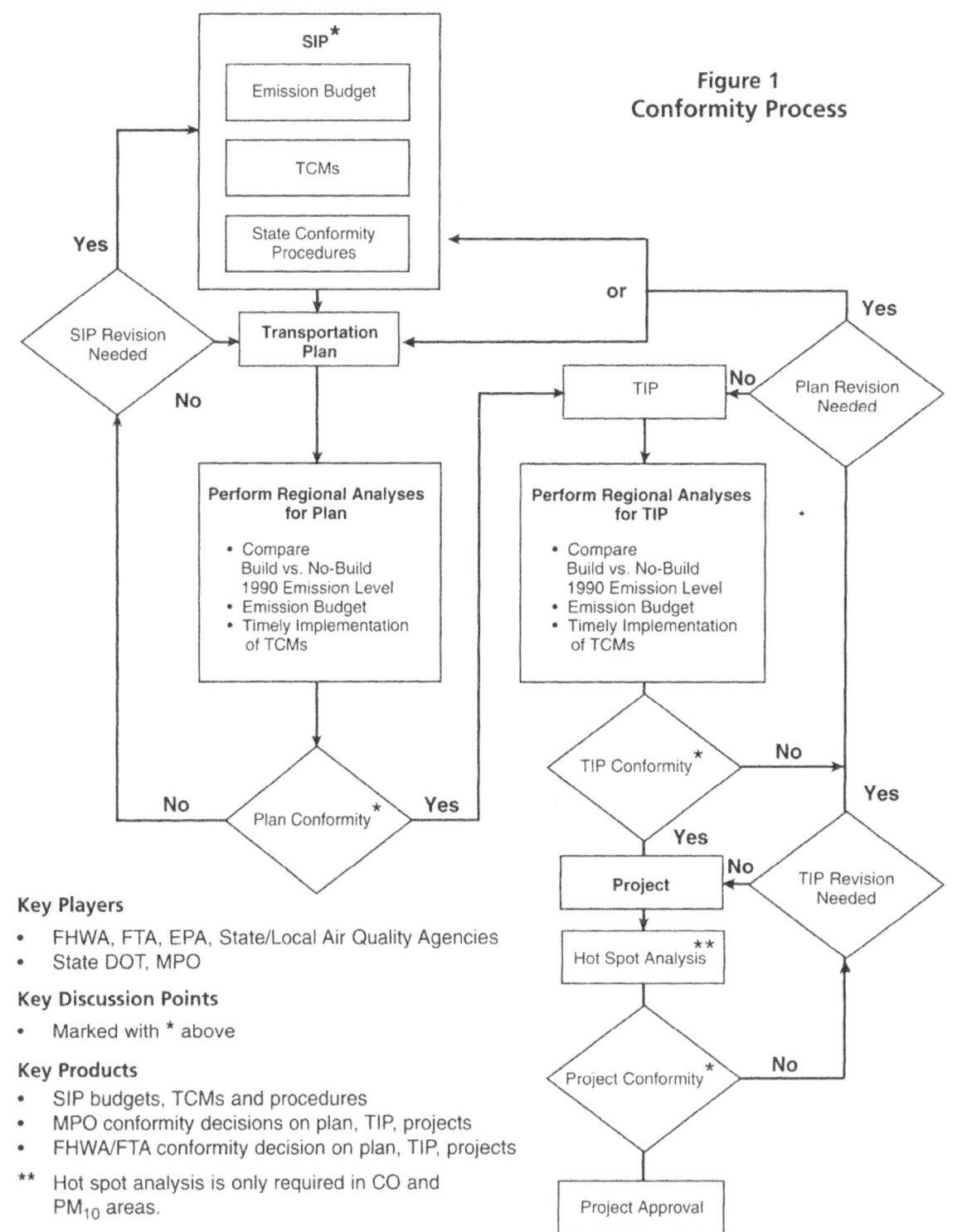

Figure 1
Conformity Process

Key Players
- FHWA, FTA, EPA, State/Local Air Quality Agencies
- State DOT, MPO

Key Discussion Points
- Marked with * above

Key Products
- SIP budgets, TCMs and procedures
- MPO conformity decisions on plan, TIP, projects
- FHWA/FTA conformity decision on plan, TIP, projects

** Hot spot analysis is only required in CO and PM_{10} areas.

Part Two: Roles and Responsibilities for the Conformity Determination

Numerous agencies are involved in the conformity process and extensive coordination and cooperation between them is necessary to comply with the conformity regulation.

This section of the guide discusses the coordination requirements, public involvement in the process, and the respective roles and responsibilities of the various entities and officials in making a conformity determination (See Figure 2).

Interagency Consultation

A formal interagency process is required in each nonattainment and maintenance area to establish procedures for consultation between MPOs, EPA, FHWA, FTA, and State and local transportation and air quality agencies.

These procedures apply to the development of the SIP, the transportation plan, the TIP, and conformity determinations. The SIP must establish interagency consultation procedures for all coordinating agencies and include specific schedules for implementation of all strategies.

Once EPA approves the element of the SIP that describes the interagency consultation process (the conformity SIP), it is enforceable by EPA as a Federal regulation. Nonattainment and maintenance areas can choose to adopt a Memorandum of Understanding (MOU) or a regulation to satisfy this requirement. Written interagency consultation procedures include general and specific processes, such as:

- Identification of the roles and responsibilities of each agency at each stage in the SIP development and transportation planning process, including technical meetings.
- A process for circulating documents (or draft documents) and supporting materials for comment before formal adoption or publication.

Public Participation: Access for Stakeholders and Citizens

ISTEA emphasizes public participation and requires proactive efforts be made to involve the public in transportation planning. These requirements also apply to the transportation conformity process. Public participation might include input to major investment studies, environmental analysis conducted under the National Environmental Policy Act, and social, economic, and environmental factors that metropolitan areas must address in their long-term planning efforts.

In short, the public must have an opportunity for early and continuing involvement in decisions. This includes citizens, affected public agencies, and other parties interested in the development of plans, TIPs, and all other elements of the planning process.

Figure 2
Roles and Responsibilities of Federal, State, and Local Transportation and Air Quality Agencies in Transportation Conformity and SIP Development Process*

(*This figure outlines general requirements and typical roles and responsibilities of the various involved agencies. Specific States and metropolitan areas may have negotiated different assignments of responsibility tailored to local conditions.)

Agencies	Action Required	When
MPO	• conduct conformity analysis on regional plan/TIP • incorporate latest emissions factors and planning assumptions • circulate draft plan/TIP for interagency and public comment • ensure public involvement procedures are developed and followed • respond to significant comments on plan/TIP conformity documents • ensure conformity of plan/TIP • ensure timely implementation of TCMs • review and comment on draft and related mobile source SIPs	• at least every three years; when a new plan, TIP, or amendments to a plan or TIP are proposed; or as needed based on SIP submittal
State Transportation Agency	Regional Analysis • conduct regional conformity analysis on projects based on interagency consultation Project Level Analysis • review and comment on draft and related mobile source SIPs • conduct "hot-spot" analysis as part of the NEPA process • provide for public involvement/respond to significant comments • ensure project level conformity • ensure timely implementation of TCMs	• when projects are proposed in CO and PM_{10} nonattainment and maintenance areas • when projects are proposed in rural nonattainment and maintenance areas. • as needed • as needed • as needed
State Air Quality/ Environmental Agency	• ensure latest emissions factors and planning assumptions are used for inventory development • prepare SIP for each relevant pollutant • hold public hearings prior to SIP adoption • prior to board approval action, ensure SIPs are complete and enforceable under the 1990 CAA • participate in interagency consultation process • forward SIP to EPA for Federal approval • ensure timely implementation of TCMs	• in accordance with CAA schedules, and as needed thereafter
State Legislature	• develop legislation to enforce applicable CAA provisions • support funding for implementation of programs	• in accordance with CAA schedules, and as needed thereafter.
U.S. DOT-FHWA/FTA	• develop technical guidance on traffic demand and forecasting, and federal-aid program guidance • provide technical guidance on TCMs and SIP development • participate in interagency consultation meetings for plan/TIP and SIP development • ensure adequate public involvement as part of the metropolitan planning process • make joint conformity determinations on MPO plans/TIPs • ensure timely implementation of TCMs	• as needed • as needed • as needed • as needed • at least every three years • for each plan/TIP conformity determination
U.S. EPA	• designate approved emissions models for use in SIP development and conformity determinations • designate "guideline" dispersion models for project level emissions analysis • participate in interagency consultation during SIP and plan/TIP development • review and comment on proposed conformity determinations • review, comment, and approve SIPs	• as needed • as needed • as needed • at least every three years • as needed

Conclusion

The CAA's transportation conformity requirement has changed the way transportation agencies develop plans and projects for funding. It has also enhanced the level of cooperation and communication between air quality and transportation agencies at all levels of government. Most importantly, the conformity requirement has been a key factor in ensuring that transportation and air quality planning are integrated at the metropolitan and State levels and that the SIP and transportation plans and programs are consistent in identifying and implementing strategies to reduce emissions from mobile sources.

State and local officials and decision makers make the key decisions on how transportation funding is expended in their States and metropolitan areas. They are also responsible for meeting National environmental objectives such as attaining the National Ambient Air Quality Standards.

At the same time, they have a responsibility to their constituents who also want improved mobility, quality of life, and economic vitality in their communities. These goals can all be addressed through a combination of good transportation planning and development of air quality plans that reflect an understanding of investment and emission reduction trade-offs and their impacts on regional, State, and National objectives.

Conformity is one tool to ensure the integration of transportation and air quality planning, and it has had an impact on funding decisions. It is anticipated that the pending amendments to the conformity regulation will help to streamline the transportation and air quality planning process.

It is important that State and local officials be fully informed about transportation and air quality choices and that they are actively engaged in key decisions. Informed decision making will enable officials to balance the important objectives of clean air, the economic well-being of communities, and improved mobility.

APPENDIX A: Options to Reduce Emissions from Motor Vehicles

The CAA identifies actions to be taken to reduce emissions from mobile sources. While some of the measures are not the responsibility of State and local transportation officials, it is beneficial for officials to be familiar with the transportation measures being taken by other public agencies (e.g., motor vehicle departments, environmental agencies), automobile manufacturers, and fuel suppliers to reduce emissions; and to understand the trade-offs between those measures and the transportation strategies and TCMs they might include in their transportation plans and programs. Having an understanding of the costs and benefits of all available options to achieve emission reductions is also useful to officials in advance of being asked to make decisions on specific strategies for implementation.

Vehicle Controls

Emission reductions resulting from the implementation of nationwide vehicle emission control strategies have been the most substantial to date in efforts to improve air quality and reduce mobile source emissions. The reductions from these strategies allowed for a doubling of vehicle miles traveled (VMT) nationwide between 1970 and 1990 while making substantial improvements in air quality. Tailpipe emission standards for cars and light-duty trucks were tightened in the CAA and, as of 1996, apply to all new vehicles sold nationwide. In addition, heavy-duty trucks have to meet new NOx emission standards after 1998. Urban transit buses also have to meet tighter emissions standards under the CAA.

Fuel-based Standards

In addition to these vehicle-related measures, stricter fuel volatility requirements (the rate at which fuel evaporates) may be required in some areas and can be useful in others. The use of reformulated gasoline (containing a different mix of ingredients than conventional gasoline) is mandated in the Nation's most serious ozone areas, although EPA now allows for an "opting-out" of the requirements under certain conditions. EPA also allows for "opting-in" to these requirements for those States that may want to use this strategy to reduce emissions. This is a measure that officials may want to consider because it is possible that all light-duty vehicles operating in the nonattainment or maintenance area could be using cleaner fuels, whereas certain other strategies apply only to limited numbers of vehicles and trips (e.g., commute trips). Oxygenated gasoline is also required during the winter months in areas with serious carbon monoxide pollution, and where low sulfur-content diesel fuel was required beginning in 1993.

The CAA includes a clean-fuel fleet program for serious ozone nonattainment areas with a population of more than 250,000 in 1980. These areas were required to adopt a clean-fuel vehicle program for centrally-fueled fleets of 10 or more vehicles by May 15, 1994. By requiring fleet owners (public and private) to convert to clean-fueled vehicles, it is hoped that the market for such vehicles will increase and broad based public acceptance of clean-fueled vehicles will increase.

An intermediate measure between mobile source controls and stationary source controls is to reduce VOC emissions by mechanical means, rather than by reducing the volatility of the fuel. In many areas, emissions from tanker truck delivery of fuel to gasoline stations are already controlled by what is known as "Stage 1" vapor recovery. This can be taken a step farther with "Stage II" vapor recovery nozzles that reduce VOC emissions from fueling individual vehicles by capturing them at the pump. Automobile manufacturers will soon begin incorporating on-board vapor recovery controls into new vehicles, but the Stage II controls could provide benefits until on-board controls are fully integrated into the vehicle fleet, in about 25 years.

Inspection and Maintenance Programs

The CAA requires inspection and maintenance programs to be adopted in certain ozone and carbon monoxide (CO) nonattainment areas. The requirements of the program vary depending upon the severity of pollution in the nonattainment areas in each State. The emission reduction potential of inspection and maintenance programs is substantial and is critical to many areas achieving the emission reductions required from mobile sources.

Transportation Control Measures

A third set of options to control and reduce emissions from motor vehicles comes under the category of TCMs. Implementation of these measures is typically within the purview of transportation agencies, and TCMs are usually funded with FHWA/FTA or State and local transportation funds. The emission reduction potential of conventional TCMs, such as ridesharing and bicycling programs is not likely to be as substantial as the transportation measures discussed above. Nevertheless, TCMs can be useful in reducing congestion and may be needed in some areas in order to demonstrate attainment of the NAAQS. TCMs, such as expanded transit services, can provide and enhance travel options and increase travel choices.

The CAA requires that for ozone nonattainment areas classified as severe or extreme, the State must identify and adopt specific transportation control strategies and TCMs to offset any projected growth in emissions from growth in vehicle miles traveled. States and MPOs should consider the CAA list of TCMs (Section 108(f)(1)(A), for strategies they might include in the SIP. These 16 TCMs (with the exception of two), also form the basis for funding eligibility in ISTEA's Congestion Mitigation and Air Quality Improvement Program (CMAQ). Below is the list of TCMs included in the CAA. There is an overlap between some of the measures, and the descriptions listed illustrate types of projects that might be considered in nonattainment areas to reduce mobile source emissions or to increase overall vehicle occupancy.

Market-Based Transportation Control Measures

In addition to conventional TCMs, work is underway in nonattainment areas to explore options to reduce mobile source emissions, using market-based TCMs such as road pricing, congestion pricing, VMT fees, and parking pricing. These mechanisms can be relatively cost-effective and *can* be designed to impact vehicles at either certain times of the day (e.g., peak-period pricing), or at all times. In addition, these measures, in combination with traditional TCMs, have the potential to address other public policy objectives such as congestion reduction and energy conservation.

Public acceptance of market-based TCMs has been slow due to practical and political considerations. Implementation of market-based measures usually requires State legislation and may require a voter referendum. Therefore, regardless of the potential merits and cost-effectiveness of these measures, the implementation of market-based TCMs is likely to occur gradually. While it is too soon to predict whether widespread use of market-based measures will occur in the future, experts generally agree that this is one option available to make substantial reductions in emissions from on-road mobile sources.

CAA Section 108(f)(1)(A) Transportation Control Measures

(i) programs for improved public transit;
(ii) restriction of certain roads or lanes to, or construction of such roads or lanes for use by passenger buses or high-occupancy vehicles (HOV);
(iii) employer-based transportation management plans, including incentives;
(iv) trip-reduction ordinances;
(v) traffic flow improvement programs that achieve emissions reductions;
(vi) fringe and transportation corridor parking facilities serving multiple-occupancy vehicle programs or transit service;
(vii) programs to limit or restrict vehicle use in downtown areas or other areas of emission concentration, particularly during periods of peak use;
(viii) programs for the provision of all forms of high-occupancy, shared-ride services;
(ix) programs to limit portions of road surfaces or certain sections of the metropolitan area to the use of non-motorized vehicles or pedestrian use, both as to time and place;
(x) programs for secure bicycle storage facilities and other facilities, including bicycle lanes, for the convenience and protection of bicyclists, in both public and private areas;
(xi) programs to control extended idling of vehicles;
(xii) reducing emissions from extreme cold-start conditions;
(xiii) employer-sponsored programs to permit flexible work schedules;
(xiv) programs and ordinances to facilitate non-automobile travel, provision and utilization of mass transit, and to generally reduce the need for single-occupant vehicle travel, as part of transportation planning and development efforts of a locality, including programs and ordinances applicable to new shopping centers, special events, and other centers of vehicle activity;
(xv) programs for new construction and major reconstruction of paths, tracks, or areas solely for use by pedestrians or non-motorized vehicles when economically feasible and in the public interest. For purposes of this clause, the Administrator shall also consult with the Secretary of the Interior;
(xvi) programs to encourage removal of pre-1980 vehicles.

APPENDIX B: Health Effects of Pollutants

EPA has established Standards for four transportation-related pollutants: ground level ozone formed by volatile organic compounds (VOCs) and oxides of nitrogen (NOx), the primary ingredients of smog ; carbon monoxide (CO); particulate matter less than 10 microns (PM_{10}); and nitrogen dioxide. The Standards are based upon EPA's assessment of the health risks associated with each of the pollutants on at-risk populations. These assessments are based upon short and long-term scientific studies by noted health professionals and medical research institutions. At-risk groups include children, the elderly, persons with respiratory illnesses, and even healthy people who exercise outdoors.

Air pollution is a phenomenon involving a complex set of chemical reactions, including combinations of pollutants, and other factors such as weather and geography. Each pollutant type plays a different role in the overall air quality in any given geographic area. Below is a brief overview of the four pollutants of most concern that must be monitored and reduced if they exceed the Standards

According to EPA, in typical urban areas, about one-third of the pollutants that create ozone (which is a combination of pollutants as discussed below) come from on-road sources such as cars, trucks, and buses. In addition, there are significant off-road sources such as construction vehicles and boats. Research indicates that large portions of the population may be at-risk due to exposure at high pollution concentrations, and under certain circumstances, health impacts can be significant. Due to concern about the health effects of air pollutants on people and the proportion of pollutants that come from cars, trucks, and buses, the provisions of the CAA relating to transportation projects, plans, and programs were made substantially more stringent than in earlier Clean Air Acts.

The principal transportation related pollutants addressed in the CAA are listed below.

Ozone and Volatile Organic Compounds (VOCs): VOCs come from vehicle exhaust, paint thinners, solvents and other petroleum-based products. VOCs and nitrogen oxides react in the presence of sunlight to form ozone. Ozone irritates the eyes, impairs the lungs, and aggravates respiratory problems. Ozone can cause chest pain, coughing, nausea, pulmonary congestion, and possible long-term lung damage. A number of exhaust VOCs are also toxic, with the potential to cause cancer.

Nitrogen Oxides (NOx): Under the high pressure and temperature conditions in an engine, nitrogen and oxygen atoms in the air react to form various nitrogen oxides, collectively known as NOx. NOx, like hydrocarbons, is a precursor to the formation of ozone and also contributes to the formation of acid rain. NOx impacts the respiratory system, causing a high incidence of acute respiratory diseases. Pre-school children are especially at risk. NOx also degrades visibility due to its brownish color and its conversion to nitrate particles.

Carbon Monoxide(CO): Carbon monoxide is a product of incomplete combustion and occurs when carbon in the fuel is partially oxidized rather than fully oxidized to carbon dioxide (CO2). Carbon monoxide reduces the flow of oxygen in the bloodstream and is particularly dangerous to persons with heart disease. Exposure to carbon monoxide impairs visual perception, manual dexterity, learning ability, and performance of complex tasks.

Particulates(PM): These are tiny particles of dust that cause irritation and damage to the respiratory system. This can result in difficulty in breathing, induce bronchitis, and aggravate existing respiratory disease. Exposure to particulates impacts individuals with chronic pulmonary or cardiovascular disease, people with influenza or asthma, children, and elderly persons. Particulates aggravate breathing difficulties, damage lung tissue, alter the body's defense against foreign materials, and can lead to premature mortality. Current standards apply to particulates less than 10 microns (PM_{10}) although a new standard for even smaller particulates is under consideration.

APPENDIX C: Resource Agencies Contact List

Below is a listing of organizations that may be contacted in order to find out what agencies are responsible for the conformity process in any given geographic area.

For State Departments of Transportation
American Association of State Highway and Transportation Officials
444 N. Capitol St. N.W.
Washington, D.C. 20001
Telephone: 202-624-5800

For Transit Agencies
American Public Transportation Association
1201 New York Avenue, N.W.
Washington, D.C. 20005
Telephone: 202-898-4000

For Metropolitan Planning Organizations or Councils of Government
National Association of Regional Councils
1700 K St. N.W.
Washington, D.C. 20006
Telephone: 202-457-0710

For State or Local Air Agencies
State and Territorial Air Pollution Program Administrators/Association of Local Air Pollution Control Officials
444 North Capitol St. N.W.
Washington, D.C. 20001
Telephone: 202-624-7864

For Federal Highway Administration Regional Offices
Either contact the FHWA Division office in each State or the FHWA Regional offices that are staffed with a Regional Air Quality Specialist.

Region	Location	Telephone
Region 1:	Albany, NY	Tel: 518-431-4236
Region 3:	Baltimore, MD	Tel: 410-962-3744
Region 4:	Atlanta, GA	Tel: 404-562-3673
Region 5:	Olympia Fields, IL	Tel: 708-283-3536
Region 6:	Fort Worth, TX	Tel: 817-978-3235
Region 7:	Kansas City, MO	Tel: 816-276-2750
Region 8:	Lakewood, CO	Tel: 303-969-6712
Region 9:	San Francisco, CA	Tel: 415-744-3823
Region 10:	Portland, OR	Tel: 503-326-2061

For Federal Transit Administration Regional Offices

Region	Location	Telephone
Region 1:	Boston, MA	Tel: 617-494-2055
Region 2:	New York, NY	Tel: 212-264-8162
Region 3:	Philadelphia, PA	Tel: 215-656-6900
Region 4:	Atlanta, GA	Tel: 404-347-3948
Region 5:	Chicago, IL	Tel: 312-353-2789
Region 6:	Arlington, TX	Tel: 817-860-9663
Region 7:	Kansas City, MO	Tel: 816-523-0204
Region 8:	Denver, CO	Tel: 303-844-3242
Region 9:	San Francisco, CA	Tel: 415-744-3133
Region 10:	Seattle, WA	Tel: 206-220-7954

For Environmental Protection Agency Contacts
Questions on transportation conformity or a current listing of nonattainment and maintenance areas should be directed to:
EPA Office of Mobile Sources
Ann Arbor, MI
Tel: 313-741-7842 or
www.epa.gov/omswww/transp.htm#conform
or
www.epa.gov\oar\oaqps\greenbk\ontc.html

GLOSSARY

Area source	Small stationary and non-transportation pollution sources that are too small and/or numerous to be included as point sources but may collectively contribute significantly to air pollution (e.g., dry cleaners).
Arterial	A class of street serving major traffic movement that is not designated as a highway.
Attainment Area	An area with air quality that meets or exceeds the U.S. Environmental Protection Agency (EPA) health standards as stated in the Clean Air Act. Nonattainment areas are areas considered not to have met these standards for designated pollutants. An area may be an attainment area for one pollutant and a nonattainment area for others.
Carbon monoxide (CO)	A colorless, odor-less, tasteless gas formed in large part by incomplete combustion of fuel. Human activities (i.e., transportation or industrial processes) are largely the source for CO contamination.
Conformity	Process to assess the compliance of any transportation plan, program, or project with air quality implementation plans. The conformity process is defined by the Clean Air Act.
Congestion Management and Air Quality Improvement Program (CMAQ)	A new categorical funding program created with the ISTEA. Directs funding to projects that contribute to meeting National air quality standards. CMAQ funds generally may not be used for projects that result in the construction of new capacity available to SOVs (single-occupant vehicles).
Emissions budget	The part of the State Implementation Plan (SIP) that identifies the allowable emissions levels, mandated by the National Ambient Air Quality Standards (NAAQS), for certain pollutants emitted from mobile, stationary, and area sources. The emissions levels are used for meeting emission reduction milestones, attainment, or maintenance demonstrations.
Emissions budget for motor vehicles	That portion of the total allowable emissions defined in a revision of the applicable State Implementation Plan (SIP), for a certain date for the purpose of meeting reasonable further progress milestones or attainment or maintenance demonstrations for any criteria pollutant or its precursors allocated by the applicable SIP to highway and transit vehicles.
Emissions inventory	A complete list of sources and amounts of pollutant emissions within a specific area and time interval.
Environmental Protection Agency (EPA)	The Federal regulatory agency responsible for administering and enforcing Federal environmental laws including the Clean Air Act, the Clean Water Act, the Endangered Species Act, and others.

Federal Highway Administration (FHWA)	An agency of the U.S. Department of Transportation that funds highway planning and programs.
Federal Transit Administration (FTA)	An agency of the U.S. Department of Transportation that funds transit planning and programs.
Inspection and Maintenance Program (I/M)	An emissions testing and inspection program implemented by states in nonattainment areas to ensure that the catalytic or other emissions control devices on in-use vehicles are properly maintained.
Intermodal	The ability to connect, and connections between, modes of transportation.
Intermodal Surface Transportation Efficiency Act of 1991 (ISTEA)	Legislative initiative by the U.S. Congress that restructured funding for transportation programs. ISTEA authorized increased levels of highway and transportation funding and an increased role for regional planning commissions/MPOs in funding decisions. The Act also requires comprehensive regional and Statewide long-term transportation plans and places an increased emphasis on public participation and transportation alternatives.
Land Use	Refers to the manner in which portions of land or the structures on them are used, e.g., commercial, residential, retail, industrial, etc.
Level of Service (LOS)	Refers to a standard measurement used by transportation officials which reflects the relative ease of traffic flow on a scale of A to F, with free-flow being rated LOS-A and congested conditions rated as LOS-F.
Long Term	In transportation planning, refers to a time span of, generally, 20 years. The transportation plan for metropolitan areas and for States should include projections for land use, population, and employment for the 20-year period.
Major Investment Study (MIS)	A study resulting from the identification of a transportation problem in a corridor or subarea that suggests the need for a major investment using Federal funds.
Metropolitan Planning Organization (MPO)	The organizational entity designated by law with lead responsibility for developing transportation plans and programs for urbanized areas with populations of 50,000 or more. MPOs are established by agreement of the Governor and units of general purpose local government which together represent 75 percent of the affected population of an urbanized area.

Mobile source	Mobile sources include motor vehicles, aircraft, seagoing vessels, and other transportation modes. The mobile source-related pollutants are carbon monoxide (CO), volatile organic compounds (VOCs), nitrogen oxides (NOx), and small particulate matter (PM_{10}).
Mobility	The ability to move or be moved from place to place.
National Ambient Air Quality Standards (NAAQS)	Federal standards that set allowable concentrations and exposure limits for various pollutants. The EPA developed the standards in response to a requirement of the CAA.
National Environmental Policy Act (NEPA)	The National Environmental Policy Act of 1969, as amended (42 U.S.C. 4321 et seq.).
National Highway System (NHS)	The Federal transportation system designated by Congress that includes nationally significant Interstate Highways and roads for interstate travel, national defense, intermodal connections, and international commerce.
Nonattainment area	A geographic region of the United States that the EPA has designated as not meeting the NAAQS.
Oxygenated gasoline	Gasoline enriched with oxygen-bearing liquids to reduce CO production by permitting more complete combustion.
Ozone (O_3)	A colorless gas with a sweet odor. Ozone is not a direct emission from transportation sources. It is a secondary pollutant formed when VOCs and NOx combine in the presence of sunlight. Ozone is associated with smog or haze conditions. Although the ozone in the upper atmosphere protects us from harmful ultraviolet rays, ground-level ozone produces an unhealthy environment in which to live. Ozone is created by human and natural sources.
Particulate matter (PM)	Any material that exists as solid or liquid in the atmosphere. Particulate matter may be in the form of fly ash, soot, dust, fog, fumes, etc.
Parts per million (ppm)	A measure of air pollutant concentrations.
Public Participation	The active and meaningful involvement of the public in the development of transportation plans and programs.
Reformulated gasoline (RFG)	Gasoline specifically developed to reduce undesirable combustion products.

Small particulate matter (PM$_{10}$)	Particulate matter which is less than 10 microns in size. A micron is one millionth of a meter. Particulate matter this size is too small to be filtered by the nose and lungs.
State Implementation Plan (SIP)	A plan mandated by the CAA that contains procedures to monitor, control, maintain, and enforce compliance with the NAAQS.
Stationary source	Relatively large, fixed sources of emissions (e.g., chemical process industries, petroleum refining and petrochemical operations, or wood processing).
Transit	Generally refers to passenger service provided to the general public along established routes with fixed or variable schedules at published fares. Related terms include: public transit, mass transit, public transportation, urban transit, and paratransit.
Transportation Control Measures (TCMs)	Actions to adjust traffic patterns or reduce vehicle use to reduce air pollutant emissions. These may include HOV lanes, provision of bicycle facilities, ridesharing, telecommuting, etc. Such actions may be included in a SIP if needed to demonstrate attainment of the NAAQS.
Transportation Improvement Program (TIP)	Also known as a transportation program, a TIP is a program of transportation projects drawn from or consistent with the transportation plan and developed pursuant to Title 23, U.S.C. (United States Code) and the Federal Transit Act.
Transportation Plan	This is a long-range plan that identifies facilities that should function as an integrated transportation system, and developed pursuant to Title 23, U.S.C. (United States Code) and the Federal Transit Act. It gives emphasis to those facilities that serve important national and regional transportation functions, and includes a financial plan that demonstrates how the long-range plan can be implemented.
U.S. Department of Transportation (DOT)	The principal direct Federal funding agency for transportation facilities and programs. Includes the Federal Highway Administration (FHWA), the Federal Transit Administration (FTA), the Federal Railroad Administration (FRA), and others.
Urbanized Area	Area which contains a city with a population of 50,000 or more plus incorporated surrounding areas meeting set size or density criteria.
Vehicle miles traveled (VMT)	The sum of distances traveled by all motor vehicles in a specified region.
Volatile Organic Compounds (VOCs)	VOCs come from vehicle exhaust, paint thinners, solvents, and other petroleum-based products. A number of exhaust VOCs are also toxic, with the potential to cause cancer.

www.ingramcontent.com/pod-product-compliance
Lightning Source LLC
Chambersburg PA
CBHW081813170526
45167CB00008B/3419